Protecting Life on Earth: Steps to Save the Ozone Layer

Cynthia Pollock Shea

Worldwatch Paper 87
December 1988

Financial support for this paper was provided by the Public Welfare Foundation.

Sections of this paper may be reproduced in magazines and newspapers with acknowledgment to the Worldwatch Institute. The views expressed in this paper are those of the author and do not necessarily represent those of the Worldwatch Institute and its directors, officers, or staff, or of funding organizations.

©Worldwatch Institute, 1988
Library of Congress Catalog Card Number 88-51743
ISBN 0-916468-88-7

Printed on Recycled Paper

Table of Contents

Introduction 5
The Ozone Depletion Puzzle 7
Effects of Ultraviolet Radiation 13
Chemical Wonders, Atmospheric Villains 18
Reducing Emissions 22
Beyond Montreal 32
Notes 38

When British scientists reported in 1985 that a hole in the ozone layer had been occurring over Antarctica each spring since 1979, the news came as a complete surprise. Although the theory that a group of widely used chemicals called chlorofluorocarbons (CFCs) would someday erode upper atmospheric ozone had been advanced in the mid-seventies, none of the models had predicted that the thinning would first be evident over the South Pole—or that it would be so severe. Scientists were baffled: What was threatening this key life-support system and how many other surprises lay in store?[1]

Ozone, the three-atom form of oxygen, is the only gas in the atmosphere that limits the amount of harmful solar ultraviolet radiation reaching the earth. Most of it is found at altitudes of between 12 and 25 kilometers, but even there, at its greatest concentration, it is present at only a few parts per million. Chemical reactions triggered by sunlight constantly replenish ozone above the tropics, and global air circulation transports some of it to the poles.

By the Antarctic spring of 1987, the average ozone concentration over the South Pole was down 50 percent. In some spots it had essentially disappeared. Although the depletion was alarming, many thought that the thinning was seasonal and unique to Antarctica, a phenomenon attributable to altered atmospheric chemistry and to perplexing polar air circulation.[2]

A scientific report issued in March 1988 shattered this view. An international group of more than 100 experts reported that the ozone

---

I sincerely thank Susan Fine and Bruce Goldstein for their research assistance and Susan Norris for production assistance. I am grateful to Douglas Cogan, Christopher Flavin, John Hoffman, Sandra Postel, Sherwood Rowland, and Michael Oppenheimer for their comments on earlier drafts of the manuscript.

layer around the entire globe was eroding much faster than models had predicted. Between 1969 and 1986, the average concentration of ozone in the stratosphere had fallen by approximately 2 percent. The magnitude of the decline varied by latitude and by season, with the most heavily populated regions of Europe, North America, and the Soviet Union suffering a year-round depletion of 3 percent and a winter loss of 4.7 percent.[3]

As ozone diminishes in the upper atmosphere, the earth receives more ultraviolet radiation, which promotes skin cancers and cataracts and depresses the human immune system. As more ultraviolet radiation penetrates the atmosphere, it will worsen these health effects, reduce crop yields and fish populations, damage some materials such as plastics, and increase smog. Continued ozone depletion will affect the well-being of every person on the planet.

Compounds containing chlorine and bromine, which are released from industrial processes and products and then waft into the upper atmosphere, are now widely accepted as the primary culprits in ozone depletion. Most of the chlorine comes from CFCs; the bromine originates from halons used in fire extinguishers.

Spurred to action by the ozone hole over Antarctica, 35 countries have signed an international agreement—the Montreal Protocol—aimed at halving most CFC emissions by 1998 and freezing halon emissions by 1992. (As of mid-November 1988, 14 nations had ratified the pact.) Although an impressive diplomatic achievement and an important first step, the agreement is so riddled with loopholes that its objectives will not be met. Furthermore, scientific findings subsequent to the negotiations reveal that even if the treaty's goals were met, significant further deterioration of the ozone layer would still occur.[4]

New evidence that global warming may already be under way strengthens the need to further control and phase out CFC and halon emissions. With their strong heat-absorbing properties, CFCs and halons are an important contributor to the greenhouse effect, accounting for 15 to 20 percent of the projected warming. Because they are synthetic chemicals for which substitutes can be developed, they are the easiest greenhouse gases to control.[5]

> "Currently available control technologies could reduce CFC and halon emissions by some 90 percent."

The fundamental links between CFCs, halons, ozone depletion, and the greenhouse effect are no longer in doubt. Currently available control technologies and stricter standards governing equipment operation and maintenance could reduce CFC and halon emissions by some 90 percent. But effective government policies and industry practices to limit and ultimately phase out chlorine and bromine emissions have yet to be formulated. Encouraging initiatives have come from some countries, particularly in Scandinavia, and from some corporations. But just as the effects of ozone depletion and climate change will be felt worldwide, a lasting remedy to the problems must also be global.

The Ozone Depletion Puzzle

In 1985, a team led by Joseph Farman, of the British Antarctic Survey, startled the world by reporting a 40-percent loss in springtime ozone over Antarctica. Using equipment developed in the twenties, the team discovered the lowest concentration of ozone ever recorded on earth. Although F. Sherwood Rowland and Mario J. Molina of the University of California at Irvine had predicted a decline in global ozone in 1974, the international scientific community was skeptical. Why hadn't state-of-the-art monitoring devices aboard sophisticated satellites detected the drop?[6]

Closer examination revealed that satellite sensors had in fact recorded the decline, and even flagged the lowest values. Scientists had simply fallen behind in processing the voluminous data. Indeed, a search of computer archives showed that in 1984 the hole had been larger than the United States and taller than Mount Everest.[7]

Faced with evidence of this local phenomenon that had not been predicted by global atmospheric models, scientists scurried to explain the cause and to determine if the depletion would spread beyond Antarctica. They advanced theories based on both chemical and natural causes and planned a reassessment of satellite and terrestrial data, along with expeditions to the continent. The first Antarctic team arrived in late August 1986, hoping to determine quickly whether the hole was caused by natural forces or by synthetic chemicals. The secret was not easily revealed. Several experiments cast doubt on the theories that blamed solar cycles and the upward movement of air from the troposphere (the layer of air surrounding the earth), but tests to confirm chemical destruction were inconclusive. Team leader Susan Solomon summed up the findings: "Based on the existing theories, chlorine is the only one we can't rule out. . . . We believe that a chemical mechanism is fundamentally responsible for the hole."[8]

Another, much larger expedition composed of 150 scientists and support personnel representing 19 organizations and four nations was launched in 1987. Based in Punta Arenas, Chile, the second National Ozone Expedition conducted satellite, aircraft, balloon, and ground measurements. Satellite data, available within 24 hours, helped researchers direct two specially outfitted planes into the center of the hole.[9]

Monitoring equipment detected that the average ozone concentration in a region twice as large as the United States dropped by nearly half from August 15 to October 7. In some areas within the hole, ozone had vanished completely. Scientists attributed the decline to a combination of factors: unique meteorological conditions in Antarctica, the presence of polar stratospheric clouds, low concentrations of nitrogen oxides, and, most important, high concentrations of active chlorine.

The pieces of the puzzle started to fall into place. During the long, sunless Antarctic winter—from about March to August—air over the continent becomes isolated in a swirling polar vortex that causes temperatures to drop below –90 degrees Celsius. This is cold enough for the scarce water vapor in the dry upper atmosphere to freeze and form polar stratospheric clouds. Chemical reactions on the surface of the ice crystals convert chlorine from nonreactive forms such as hydrogen chloride and chlorine nitrate into molecules that are very sensitive to sunlight. Gaseous nitrogen oxides ordinarily able to inactivate chlorine are transformed into frozen, and therefore nonreactive, nitric acid.[10]

Spring sunlight releases the chlorine, starting a virulent ozone-destroying chain reaction that proceeds unimpeded for five or six weeks. Molecules of ozone are transformed into molecules of ordinary,

> "Throughout the hole, increased levels of chlorine monoxide correlated with decreased concentrations of ozone."

two-atom oxygen. The chlorine emerges unscathed, ready to attack more ozone. Experiments conducted by James Anderson and his colleagues at Harvard University found that levels of chlorine monoxide—an intermediate product of the chemical reaction—were up to 500 times higher than normal in the most disturbed regions. Throughout the hole, increased levels of chlorine monoxide correlated with decreased concentrations of ozone. (See Figure 1.)[11]

Diminished ozone in the vortex means the atmosphere there absorbs less incoming solar radiation, thereby perpetuating lower temperatures and the vortex itself. In 1987, the vortex did not break up until early December, a month later than usual. In addition, the hole was some 8 degrees Celsius colder at an altitude of 15 kilometers than it was in 1979. As a result, polar stratospheric clouds were more pervasive and persistent. In essence, the ozone hole feeds on itself.[12]

Paradoxically, the phenomenon of global warming encourages the process. Higher concentrations of greenhouse gases are thought to be responsible for an increase in the earth's surface temperature and a decrease in the temperature of the stratosphere. In addition, methane, one of the primary greenhouse gases, is a significant source of stratospheric water vapor. Colder temperatures and increased moisture both facilitate the formation of stratospheric clouds.[13]

Since the ozone hole cannot get much deeper, some fear it may spread outward, encompassing larger areas of Argentina and Chile and expanding above portions of Australia, Brazil, New Zealand, and Uruguay. These areas may also suffer when the vortex breaks up and its ozone depleted air diffuses throughout the lower Southern Hemisphere. In December 1987, three of the five ozone-monitoring stations in Australia observed a sharp drop in ozone. The abnormally low values persisted for three weeks over Melbourne, resulting in the lowest December mean ozone levels on record. Roger Atkinson of the Australian Bureau of Meteorology wonders if "it's possible that this is the first sign of depletion extending over Australia."[14]

While many of the meteorological and chemical conditions conducive to ozone depletion are unique to Antarctica, ground-based research in Greenland in the winter of 1988 found elevated chlorine concentra-

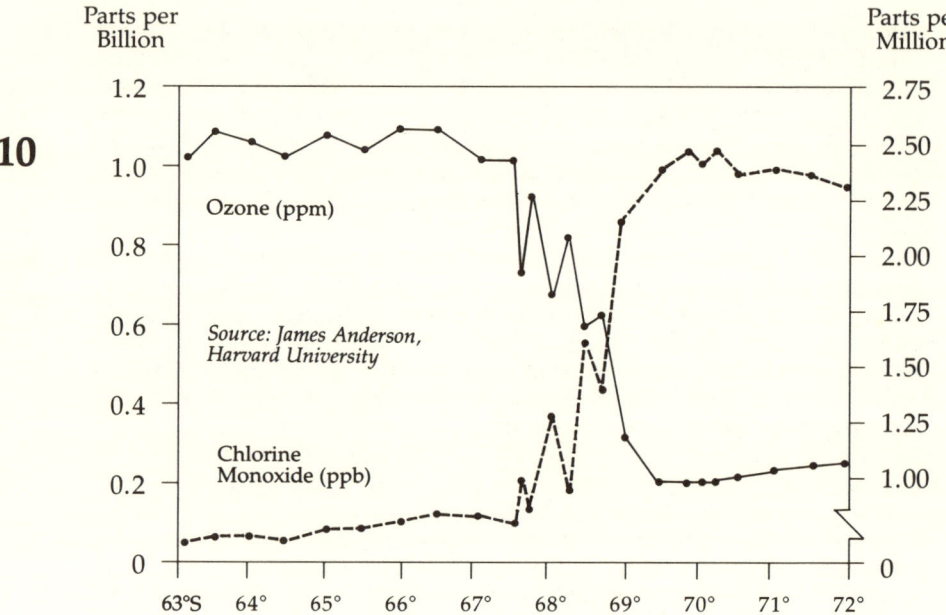

Figure 1: Atmospheric Concentration of Chlorine Monoxide and Ozone by Latitude, Southern Hemisphere, 1987

tions and depressed ozone levels over the Arctic as well. Although a strong vortex does not develop there and temperatures are not as low, polar stratospheric clouds do form.[15]

To find out more about the atmospheric chemistry of the Arctic, the U.S. National Aeronautics and Space Administration (NASA) will send teams to Stavanger, Norway, and Ellesmere Island, Canada, in early 1989. The expeditions will use instruments on the ground to analyze the chemical components of the atmosphere, and similar instruments aboard airplanes to sample the air in the ice clouds. Meanwhile, sensors aboard the Nimbus-7 satellite will examine reflected sunlight. Ground-based measurements will be gathered simultaneously by the Danes in Greenland and by the Soviets over their

> "Ozone loss has been documented around the globe, not just at the poles, and is occurring far more rapidly than had been forecast."

northernmost territories. The Soviet Union also plans aircraft observations at about the same time.[16]

The theories on how chlorine interacts on the surface of particles in polar stratospheric clouds are leading to worries that similar ozone-depleting reactions may occur around the globe. The importance of cloud surface chemistry in the destruction of ozone was not realized until 1986. Now, its role is viewed as fundamental. Some scientists believe that sulfate chemical surfaces may host the same catalytic chlorine reactions that occur over Antarctica. Sulfate aerosols from volcanoes and biological processes are present in the atmosphere at all latitudes, at heights of 15 to 22 kilometers. They are especially prevalent over the most densely populated areas of the Northern Hemisphere, perhaps indicating the involvement of sulfur emissions from human activities. If these chemicals are capable of hosting the same catalytic reactions, global ozone depletion may accelerate even more rapidly than anticipated.[17]

Consensus about the extent of ozone depletion and its causes strengthened with the release of the NASA Ozone Trends Panel report on March 15, 1988. More than 100 scientists from 10 countries spent 16 months reviewing the published literature and performing a critical reanalysis and interpretation of nearly all ground-based and satellite-derived ozone data. Their purpose: to eliminate any errors caused by improperly calibrated instruments.

Ozone losses were documented around the globe, not just at the poles. The blame, particularly for the Antarctic ozone hole, was firmly placed on chlorofluorocarbons. The panel reported that between 30–64 degrees north latitude, where most of the world's people live, the total amount of ozone above any particular point had decreased by between 1.7 and 3.0 percent in the period from 1969 to 1986. (See Table 1.) Further, "ozone decreases were most pronounced in winter, ranging from 2.3 to 6.2 percent (depending on latitude), and those winter changes were higher than predicted by atmospheric models."[18]

Because monitoring stations are not as prevalent in the Southern Hemisphere, the panel cautioned that data for regions south of 30 degrees north were not as reliable, nor could seasonal variations be

Table 1: Global Decline in Atmospheric Ozone, 1969–86[1]

Latitude	Year Round Decrease (percent)	Winter Decrease (percent)
53–64° N	− 2.3	− 6.2
40–53° N	− 3.0	− 4.7
30–40° N	− 1.7	− 2.3
19–30° N	− 3.1	n.a.
0–19° N	− 1.6	n.a.
0–19° S	− 2.1	n.a.
19–29° S	− 2.6	n.a.
29–39° S	− 2.7	n.a.
39–53° S	− 4.9	n.a.
53–60° S	−10.6	n.a.
60–90° S	− 5.0 or more	n.a.

[1]Data for the area 30 to 64 degrees north of the equator are based on information gathered from satellites and ground stations from 1969 to 1986. Data for the area from 60 degrees south to the South Pole are based on information gathered from satellites and ground stations since 1979. All other information was compiled after November 1978 from satellite data alone.

Sources: NASA Ozone Trends Panel; Cass Peterson, "Evidence of Ozone Depletion Found Over Big Urban Areas," *Washington Post*, March 16, 1988.

accurately determined. The report further stated that while the problem was worst over Antarctica during the spring, "ozone appears to have decreased since 1979 by 5 percent or more at all latitudes south of 60 degrees south throughout the year." The hole alone covers approximately 10 percent of the Southern Hemisphere.[19]

The report's findings startled policymakers, industry representatives, and researchers around the world. Prior to the NASA report, the phenomenon of global ozone depletion and the role of CFCs had been hotly contested. Within a matter of weeks the report's conclusions

> "Erosion of the ozone shield could result in 5-20 percent more ultraviolet light reaching population areas within the next 40 years."

were widely accepted, and public debate on the issue began to build. Suddenly, ozone depletion was real, no longer just a theory, and people around the globe knew just how bad the problem overhead had become.

Scientists were alarmed not only by the documented damage to the ozone layer, but by the inadequacy of their models to predict it. Ozone depletion is occurring far more rapidly and in a different pattern than had been forecast. Projections of the amount and location of future ozone depletion are still highly uncertain. Although the fundamental mechanisms of ozone depletion are generally understood, the quantitative effect of cloud surface chemistry, the rate of various chemical reactions, and the specific chemical pathways are still in doubt. According to Sherwood Rowland, one of the first to sound a warning, policy decisions now and for at least another decade must be made without good quantitative guidelines of what the future holds.[20]

Effects of Ultraviolet Radiation

Tanned, wrinkled, and leathered skin, certain eye disorders, and brittle plastics are all caused by ultraviolet radiation that reaches the earth's surface. At present, ozone absorbs much of the ultraviolet light that the sun emits in wavelengths harmful to humans, animals, and plants. (The most biologically damaging wavelengths are within the 290–320 nanometer band, referred to as UV-B.) But, according to uncertain projections from computer models, erosion of the ozone shield could result in 5–20 percent more ultraviolet radiation reaching populated areas within the next 40 years—most of it in the UV-B band.[21]

In light of the findings of the NASA Ozone Trends Panel, the U.S. Environmental Protection Agency (EPA) damage projections cited in this section are conservative. Although the EPA ranges are based on current control strategies, they assume ozone depletion levels of 1.2–6.2 percent. Yet, all areas of the globe have already suffered depletion beyond this lower bound. At very high latitudes, depletion has exceeded the EPA's most realistic upper bound. Thus although these damage estimates are the best currently available, they are probably on the low side.[22]

Globally, skin cancer incidence among Caucasians is already on the rise, largely because of more outdoor activity, but it is expected to increase alarmingly in the presence of more UV-B. Some 600,000 new cases of squamous and basal cell carcinoma—the two most common but rarely fatal skin cancer types—are reported each year in the United States alone. Worldwide, the number of cases is at least three times as high. The incidence is closely tied to cumulative exposure to ultraviolet radiation.[23]

Each 1-percent drop in ozone is projected to result in 4–6 percent more cases of these types of skin cancer. Ozone depletion is expected to cause 3 million to 15 million new cases in Americans born before 2075; some 52,000 to 252,000 of those patients are likely to die from the disease. Most at risk are those with light coloring who live nearest the equator. Also particularly susceptible are Argentineans, Australians, Chileans, and New Zealanders who live in areas under the springtime reach of the Antarctic hole. Members of the U.S. National Science Foundation are working with colleagues in Argentina and Chile to measure the increased amount of incoming ultraviolet radiation.[24]

Melanoma, a more deadly form of skin cancer, already attacks 26,000 Americans annually and results in 8,000 deaths. Melanoma represents only 4 percent of all skin cancers but is responsible for 65 percent of all skin cancer deaths. Although the tie between melanoma and increased UV-B levels is less clear, the EPA estimates that ozone depletion will lead to an additional 31,000 to 126,000 cases among U.S. whites born before 2075, resulting in an additional 7,000 to 30,000 fatalities.[25]

Melanoma incidence and mortality are already rising in all Caucasian populations studied. In Australia, melanoma deaths have risen fivefold over the past 50 years. In the United States, the number of cases has increased 83 percent over the past seven years. In other countries, the incidence is increasing at 3–7 percent annually. Unlike basal and squamous cell carcinoma, melanoma appears to be associated with acute radiation exposure, such as severe sunburns.[26]

Under the same EPA scenarios, from 555,000 to 2.8 million Americans born before 2075 will suffer from cataracts who would not have otherwise. Victims will also be stricken earlier in life, making treatment more difficult. Cataracts, a clouding of the lens that blurs vision, can be removed in a relatively simple operation, but if left untreated can result in blindness. The surgical procedure is performed routinely in industrial countries, but victims in the developing world are less likely to have access to, or funds for, the operation.[27]

Medical researchers also fear that UV-B depresses the human immune system, lowering the body's resistance to attacking micro-organisms, making it less able to fight the development of tumors, and rendering it more prone to infectious diseases, such as herpes. In developing countries, particularly those near the equator that are exposed to higher UV-B levels, parasitic infections could become more common. The response may even decrease the effectiveness of some inoculation programs, such as those for diphtheria and tuberculosis. Instead of building an immune response to the antigen, the patient might develop a full-fledged case of the disease.[28]

Unlike skin cancer, which predominantly affects whites, a lowered immune response is likely to affect everyone. Individuals already immunosuppressed, such as transplant patients or those with acquired immunodeficiency syndrome (AIDS), could be at greater risk due to additive effects. Although the extent of immunosuppression cannot yet be quantified, some evidence suggests it could be induced with much lower doses of ultraviolet radiation than those required to cause cancer. This may mean that doses too low to cause a sunburn could lower the body's resistance to disease. An Australian study has already measured immunological effects in sunbathers.[29]

Terrestrial and aquatic ecosystems are also affected. Screenings of more than 200 different plant species, most of them crops, found that 70 percent were sensitive to UV-B. Increased exposure to radiation may decrease photosynthesis, water-use efficiency, yield, and leaf area. Soybeans, a versatile and protein-rich crop, are particularly susceptible. Alan Teramura, at the University of Maryland, discovered that a simulated ozone loss of 25 percent reduced the yield of one important soybean species by as much as 25 percent. He also found that plant sensitivity to UV-B increased as the phosphorus level in the soil increased, indicating that heavily fertilized agricultural areas may be the most vulnerable.[30]

Although plants can be bred for many traits, including ultraviolet-tolerance, rapid ozone depletion could overwhelm the capacity to evolve protective mechanisms. As a consequence, those varieties most resistant to ultraviolet light would be the ones to thrive, not necessarily those with the greatest economic value or nutritional content. So far, only 4 of the 10 major types of terrestrial ecosystems have been examined; tropical forests, rangelands, and wetlands are among those yet to be studied.[31]

Aquatic ecosystems may be the most threatened of all. Phytoplankton, the one-celled microscopic organisms that engage in photosynthesis while drifting on the ocean's surface, are the backbone of the marine food web. Because they require sunlight, already the limiting growth factor in high-latitude ocean areas, they cannot escape incoming ultraviolet radiation and continue to thrive. Yet if they remain at the water's surface, studies show that a 25-percent reduction in ozone would decrease their productivity by about 35 percent.[32]

Experiments with marine diatoms, a minute algae, show reductions in biomass, protein, and chlorophyll at UV-B levels corresponding to ozone reductions of 5–15 percent. A significant destruction of phytoplankton and its subsequent decomposition could even raise carbon dioxide levels, speeding the warming of the atmosphere.[33]

Zooplankton and the larvae of several important fish species will be doubly strained: they too live on the water's surface and their sole food supply, phytoplankton, will be scarcer. Preliminary studies indicate that sunlight is essential to the larval stage of growth, yet there is no adaptation or flight mechanism to respond to increased ultraviolet radiation. For some shellfish species, a 10-percent decrease in ozone could result in up to an 18-percent increase in the number of abnormal larvae. A study of anchovies calculates that 8 percent of the annual larvae population would be killed by a 9-percent decrease in ozone.[34]

> "Exposure resulting from ozone loss of 10 percent corresponds to moving 30 degrees closer to the equator—like moving New York City to Caracas, Venezuela."

Commercial fish populations already threatened by overharvesting may have more difficulty rebuilding due to effects of increased UV-B. Most worrisome to many marine biologists is how the species composition of ocean environments will change. Some species will undoubtedly be more vulnerable to increased ultraviolet radiation than others and the changes are likely to be dramatic. Exposure resulting from ozone loss of 10 percent corresponds to moving 30 degrees closer to the equator—like moving New York City to Caracas, Venezuela. Ultimately, entire ecosystems may become more unstable and less flexible.

Increased UV-B levels also affect synthetic materials. Plastics are especially vulnerable. Studies conducted for the EPA estimated that without added chemical stabilizers, the cumulative damage to just one polymer, polyvinyl chloride, could reach $4.7 billion by 2075 in the United States alone.[35]

Ozone, the same substance that acts as a protective shield in the stratosphere, is a dangerous pollutant at ground level. Ironically, destruction of the upper ozone could increase the amount at the earth's surface. As more ultraviolet radiation reaches the ground, the photochemical process that creates smog will accelerate. Studies show that ground-level ozone, the main component of smog, retards crop and tree growth, limits visibility, and impairs lung functions. Urban air quality, already poor in most areas of the world, will worsen. In addition, stratospheric ozone decline is predicted to increase tropospheric amounts of hydrogen peroxide, an acid rain precursor.[36]

Most of the research to date on the effects of increased ultraviolet radiation has been conducted by U.S. government agencies. Their annual funding to investigate the many outstanding questions was recently boosted to some $15–20 million. West Germany has its own UV-B program, financed by the Ministry of Research and Technology, and joint programs under the auspices of the European Community are to commence shortly. Ironically, the most uncertainty surrounds the effects of increased UV-B on the immune system and on aquatic and plant life, where there is the greatest potential to harm our health and food supplies.[37]

Despite the many uncertainties regarding the amount of future ozone depletion, rising UV-B levels, and their biological effects, it is clear that the risks to aquatic and terrestrial ecosystems and to human health are enormous. The central conclusion of the EPA studies is that "the benefits of limiting future CFC/halon use far outweigh the increased costs these regulations would impose on the economy." In the United States alone, the present value of the benefits of controlling emissions through the year 2075 is estimated at $6 trillion—some 240 times greater than the costs.[38]

Chemical Wonders, Atmospheric Villains

Chlorofluorocarbons are remarkable chemicals. They are neither toxic nor flammable at ground level, as demonstrated by their discoverer, Thomas Midgley Jr., in 1930 when he inhaled vapors from a beaker of clear liquid and then exhaled to extinguish a candle. A safe coolant that was inexpensive to produce was exactly what the refrigeration industry needed. E.I. du Pont de Nemours & Company marketed the compound under the trademark Freon. (In chemical shorthand, it is referred to as CFC-12.) International production soared, rising from 545 tons in 1931 to 20,000 tons in 1945.[39]

Another use for the chemical, as a blowing agent in rigid insulation foams, was discovered in the late forties. In this application, liquid CFC-12 is vaporized into a gas that forms lightweight, closed-cell bubbles that are poor conductors of both heat and cold. Styrofoam, the most widely known of these products, was developed and marketed by the Dow Chemical Company. Total CFC-12 production doubled in the five years after 1945.[40]

Over time, new chemical formulations were discovered, and the versatility of the various CFCs seemed almost endless. CFC-11 and CFC-12 were first used as aerosol propellants during World War II in the fight against malaria. In the postwar economy, they were employed in aerosol products ranging from hairspray and deodorant to furniture polish. By the late fifties, a combination of the blowing agents CFC-11 and carbon dioxide was used to make softer furniture cushions, carpet padding, and automobile seats.

> "Many social and technological developments in recent decades were assisted, at least in part, by the availability of CFCs."

Many social and technological developments in recent decades were assisted, at least in part, by the availability of CFCs. Huge air conditioners made it possible to build and cool shopping malls, sports arenas, and high-rise office buildings. Artificial cooling brought comfort, business, and new residents to regions with warm climates. Automobile air conditioners, now installed in 80 percent of the cars sold in the United States, put the nation on wheels for summer vacations. And healthier, more interesting diets are now available because three-fourths of the food eaten in the United States is refrigerated at some point in the production and distribution chain.[41]

Even the computer revolution was aided by CFCs. As microchips and other components of electronic equipment became smaller and more sophisticated, the need to remove the smallest contaminants became critical. CFC-113 is used as a solvent to remove glue, grease, and soldering residues, leaving a clean, dry surface. And, as is the case with so many of the CFCs, the chemical is versatile enough to be used on metals and plastics and to dry-clean fabrics. CFC-113 is now the fastest growing member of the CFC family; worldwide production exceeds 160,000 tons per year.[42]

Following the energy crises of the seventies, industrial, commercial, and residential customers looked for new ways to trim their heating and electric bills. Demand for rigid-foam insulation, some types blown with CFC-11 and others with CFC-12, soared. In 1985, rigid foam accounted for two-thirds of the insulation installed in new U.S. commercial buildings, half the insulation in new single-family houses, and one-third of the home re-insulation market. Appliance manufacturers also set to work—and are now required by U.S. law—to boost the energy efficiency of their products. Today, the foam walls of a household refrigerator contain five times as much CFC (1 kilogram) as is used for a refrigerant.[43]

An industry-sponsored group, the Alliance for Responsible CFC Policy, pegs the market value of CFCs produced in the United States at $750 million annually, the value of goods and services directly dependent on the chemicals at $28 billion, and the end-use value of installed equipment and products at $135 billion. Products that currently depend on the chemicals are ubiquitous: the United States alone has 85 million household refrigerators and 60 million auto air conditioners. In addition, billions of foam packaging containers are produced each year to keep fast foods hot, eggs cushioned, and meats attractively displayed.[44]

Around the world, aerosols are still the largest user of CFCs, accounting for 25 percent of the total. (See Table 2.) Rigid foam and solvent applications, the fastest growing uses for CFCs, are tied for second place.

In 1987, global CFC production (excluding China, the Soviet Union, and Eastern Europe) surpassed the peak set in 1974, and came close to 1 million tons. Combined production of CFC-11 and CFC-12 accounts for at least three-fourths of this total. Only some two dozen firms make CFCs, but production data are scarce and not totally reliable. Companies claim that such numbers are proprietary and that having them publicly available would put the manufacturers at a competitive disadvantage. This lack of information makes it difficult to analyze and regulate CFC markets.[45]

Total per capita use of the three most common CFCs is highest in the United States—at 1.22 kilograms—but Europe and Japan are not far behind. (See Table 3.) In most of the rest of the world, consumption rates are far lower. Indeed, Americans use six times the global average. Consumption patterns also differ by region: Europeans use the most CFC-11, Americans the most CFC-12, and Japanese the most CFC-113. This is because aerosols account for 37 percent of European CFC use, mobile air conditioning and other refrigeration constitute 39 percent of use in the United States, and solvents, primarily for the electronics industry, constitute 39 percent of Japanese CFC use.[46]

From 1931 through 1986, virtually all the CFC-11 and CFC-12 produced was sold to customers in the Northern Hemisphere. Since raw chemicals and products made with and containing CFCs were then exported, in part to developing countries, final usage was not quite as lopsided. Indeed, the Third World accounted for 16 percent of global CFC consumption in 1986. (See Table 4.) As populations, incomes, and the manufacturing base grow in developing countries, CFC use there is projected to rise.[47]

Table 2: Global CFC Use, by Category, 1985

Use	Share of Total
	(percent)
Aerosols	25
Rigid-foam Insulation	19
Solvents	19
Air Conditioning	12
Refrigerants	8
Flexible Foam	7
Other	10

Source: Daniel F. Kohler et al., *Projections of Consumption of Products Using Chlorofluorocarbons in Developing Countries* (Santa Monica, California: Rand Corporation, 1987).

Unlike most chemicals, CFCs are not broken down in the troposphere. Instead, they waft slowly upward and after six to eight years reach the stratosphere. Once there, the chemicals can survive for up to 100 years. When they are broken down, each chlorine atom released is capable of destroying tens of thousands of ozone molecules before it eventually gets washed out of the atmosphere.

Halons are also inert at ground level. They contain bromine, a more effective ozone destroyer than chlorine, and are also long-lived in the atmosphere. Halons are used for fighting fires: halon 1211 is employed in hand-held extinguishers and halon 1301 is used in total-flooding systems designed for enclosed areas with valuable contents, such as computer rooms, telephone exchanges, museums, and bank storage vaults.

Halons were developed by the U.S. Army Corps of Engineers at the end of World War II as a way to fight fires in tanks and armored personnel carriers. Because they are nontoxic and can be applied directly to sensitive equipment without causing damage or leaving a residue, they have become the favored chemical for fighting fires.

Demand quadrupled between 1973 and 1984 and is still growing at a rate of 15 percent annually.[48]

Alarming though the latest ozone measurements are, they reflect only the response to gases released through the early eighties. Gases now rising through the lower atmosphere will take up to eight years to reach the stratosphere. And, an additional 2 million tons of substances containing chlorine and bromine are still on the ground, trapped in insulation foams, appliances, and fire-fighting equipment.[49]

Chlorine concentrations in the upper atmosphere have grown from 0.6 to 2.7 parts per billion in the past 25 years. Under even the most optimistic regulatory scenarios, they are expected to triple by 2075. Bromine concentrations are projected to grow considerably faster — exhibiting a tenfold increase from their current 1 part per trillion level despite an anticipated freeze on consumption.[50]

CFCs and halons are insidiously and inexorably destroying the planet's protective ozone shield. Biological systems around the globe will soon start to suffer adverse effects. Without a complete and rapid phaseout of CFC and halon production, the real losers will be future generations who inherit an impoverished environment and wonder at the folly of their ancestors.

Reducing Emissions

On September 16, 1987, after years of arduous and heated negotiation, the Montreal Protocol on Substances that Deplete the Ozone Layer was signed by 24 countries. Provisions of the agreement include a freeze on CFC production (at 1986 levels) by 1989, a 20-percent decrease in production by 1993, and another 30-percent cut by 1998. Halon production is subject to a freeze based on 1986 levels starting in 1992.[51]

In order to take effect by the target date of January 1989, the protocol must be ratified by 11 countries, representing at least two-thirds of international CFC consumption. By mid-November 1988, 14 countries had already approved the treaty—Canada, Egypt, Japan, Kenya,

Table 3: Per Capita Use of CFC-11, CFC-12, and CFC-113, 1986

	CFC-11	CFC-12	CFC-113	Total[1]
		(kg/capita)		
United States	.34	.58	.31	1.22
Europe	.47	.34	.12	.93
Japan	.23	.29	.43	.91

[1] Rows not completely additive due to trade.

Source: U.S. Environmental Protection Agency, *Regulatory Impact Analysis: Protection of Stratospheric Ozone* (Washington, D.C., 1987).

Luxembourg, Mexico, New Zealand, Nigeria, Norway, Portugal, the Soviet Union, Sweden, Uganda, and the United States. But the two-thirds consumption requirement will not be met until the treaty is ratified by the entire European Community—a step the group has pledged to take.

The means to achieve these reductions are left to the discretion of individual nations. Most signatory countries are responding with production limits on chemical manufacturers. Although this approach complies with treaty guidelines, it effectively ensures that only the highest bidders will be able to continue using CFCs. It also places the onus of curbing emissions on the myriad industrial users of the chemicals and on the consumers of products which incorporate them.

Moving quickly to protect the ozone layer calls for a different approach—one that targets the largest sources of the most ozone depleting chemicals. To do this requires knowing how much of each chemical is currently used, its emissions profile, and the uses to which it is put. (See Table 5.) Only then can individual countries assess the technical and economic feasibility of limiting emissions from specific markets.

Immediate reductions in CFC emissions can be achieved by banning CFC propellants in aerosols and by eliminating rapid evaporation of cleaning solvents. Intermediate savings are obtainable by capturing the blowing agents used to inflate flexible foams, by plugging the leaks in refrigeration and air conditioning systems, and by recovering the refrigerants drained when cooling systems are serviced or repaired. Long-term reductions call for alternative product-disposal methods, use of substitute chemicals, and development of technologies that do not rely upon ozone depleting substances.

When concern about the ozone layer first emerged in the seventies, some industrial-country governments responded. Since 56 percent of combined CFC-11 and CFC-12 production in 1974 was used in aerosols, spray cans were an obvious target. Under strong public pressure, Canada, Norway, Sweden, and the United States banned CFC propellants in at least 90 percent of their aerosol products. The change brought economic as well as environmental benefits. Hydrocarbons, the replacement propellant, are less expensive than CFCs and saved the U.S. economy $165 million in 1983 alone. The European Community adopted a different approach. In 1980, the member countries agreed not to increase their already excessive capacity to produce these two CFCs and called for a 30-percent reduction in their use in aerosol propellants by 1982 (based on 1976 consumption figures).[52]

Cumulative reductions in CFC-11 and CFC-12 emissions due to lowered use in aerosols in the United States and the European Community amount to 2 million and 501,000 tons, respectively, the equivalent of six years of current CFC-11 production and one year of CFC-12 output. These figures are based on the assumption that CFC use in aerosols would have remained at the 1974 peak in the United States and the 1976 peak in the European Community.[53]

Worldwide, aerosol cans are still the largest source of CFC emissions, contributing 224,000 tons annually—some 33 percent of combined global production of CFC-11 and CFC-12. Rising concern about ozone depletion among consumers and governments should soon curtail this use. And because some nations took the lead, economical and often less expensive substitutes are already widely available.[54]

Table 4: CFC Consumption by Region, 1986

Region	Share of Total (percent)
United States	29
Other Industrial Countries[1]	41
Soviet Union, Eastern Europe	14
Other Developing Countries	14
China and India	2

[1]European Community accounts for more than half, followed by Japan, Canada, Australia, and others.

Source: "The Ozone Treaty: A Triumph for All," *Update from State*, May/June 1988.

Denmark banned aerosol propellants in 1987, and voluntary industry cutbacks of 90 percent by the end of 1989 were announced recently in Belgium, the Netherlands, Switzerland, the United Kingdom, and West Germany. British and Swiss manufacturers will also label their products so consumers will know if they are ozone friendly. The Soviet Union has declared its intention to switch to non-CFC aerosol propellants by 1993. France, Japan, and the nations of Eastern Europe and the Third World are the only major users that have not yet announced control measures.[55]

Despite rapid growth, CFC-113 emissions may be some of the easiest and most economical to control. The chemical is only used to clean the final product and is not incorporated in it. Thus emissions are virtually immediate; three-fourths result from vapor losses, the remainder from waste disposal. A U.S. ban on land disposal of chlorinated solvents that took effect in November 1986, consideration of similar regulations elsewhere, the high cost of incinerating CFC-113 (because it contains toxic fluorine), and accelerating concern about ozone depletion have all created strong incentives for solvent recovery and recycling.[56]

Since CFC-113 costs about twice as much as other CFCs, investments in recovery and recycling pay off more quickly. Recycling of CFC-113 is now practiced on-site at many large computer companies. An IBM plant near Stuttgart, West Germany, has installed a recycling system that recovers 70–90 percent of the plant's solvents. Similar rates are being achieved by AT&T in the United States. Smaller electronics firms, for whom in-house recycling is not economical, can sell their used solvents to commercial recyclers or the distributors of some chemical manufacturers.[57]

Rapid progress in emissions reductions over the past several years bodes well for more short-term savings. Hirotoshi Goto, director of the Stratospheric Protection Program in Japan, expects industries that use CFC solvents to achieve recycling rates of 95 percent. Such firms account for some 40 percent of total Japanese CFC consumption.[58]

Capturing CFC emissions from flexible-foam manufacturing can also be accomplished fairly quickly, but requires investment in new ventilation systems. (When rigid foams are made, 90 percent of the CFC blowing agent remains in the closed cells of the product.) Current production processes for flexible foams result in the complete and immediate release of the blowing agent to the atmosphere. But new suction systems coupled with carbon adsorption technologies are able to recover from 40 to 90 percent of the CFCs released.[59]

One technology operating in both Denmark and Norway traps CFCs at the blowing stage and recovers 40–45 percent of total emissions. A more comprehensive system designed by Hyman Development in the United Kingdom is able to recover almost twice as much. Traditionally, flexible foam is cured in an open room for several days, allowing CFCs to waft from the product. In the Hyman process, curing time is reduced to 40 minutes and occurs in an enclosed area. This allows the CFCs to be captured by the ventilating system. Unifoam, a Swiss company, is marketing a similar system able to recover 85 percent of the blowing agent for future use.[60]

Another area that offers significant savings, at a low cost, is improved design, operating, and maintenance standards for refrigeration and air conditioning equipment. These uses account for 30 percent of com-

Table 5: Use and Emissions Profiles of Commonly Used Chemicals, 1985

Chemical	Emissions (thousand tons)	Atmospheric Lifetime[1] (years)	Applications	Annual Growth Rate (percent)	Share of Contribution to Depletion[2] (percent)
CFC-12	412	139	Air Conditioning, Refrigeration, Aerosols, Foams	5	45
CFC-11	238	76	Foams, Aerosols, Refrigeration	5	26
CFC-113	138	92	Solvents	10	12
Carbon Tetrachloride	66	67	Solvents	1	8
Methyl Chloroform	474	8	Solvents	7	5
Halon 1301	3	101	Fire Extinguishers	n.a.	4
Halon 1211	3	12	Fire Extinguishers	23	1
HCFC-22	72	22	Refrigeration, Foams	11	0

[1] Time it takes for 63 percent of the chemical to be washed out of the atmosphere.
[2] Total does not add to 100 due to rounding. Contribution of HCFC-22 rounds to zero.

Sources: James K. Hammitt et al., "Future Emission Scenarios for Chemicals that May Deplete Stratospheric Ozone," *Nature*, December 24, 1987; U.S. Environmental Protection Agency, *Regulatory Impact Analysis: Protection of Stratospheric Ozone*, Volume II, Part I (Washington, D.C.: 1987); Douglas Cogan, *Stones in a Glass House: CFCs and Ozone Depletion* (Washington, D.C.: Investor Responsibility Research Center, 1988).

bined CFC-11 and CFC-12 consumption. Codes of practice to govern equipment handling are being drawn up by many major trade associations. Key among the recommendations are to require worker training, to limit maintenance and repair work to authorized personnel, to install leak detection systems, and to use smaller refrigerant charges. Another recommendation, to prohibit venting of the refrigerant directly to the atmosphere, requires the use of recovery and recycling technologies.

Careful study of the automobile air conditioning market in the United States, the largest user of CFC-12 in the country, has found that 34 percent of emissions can be traced to leakage, 48 percent occur during recharge and repair servicing, and the remainder happen through accidents, disposal, and manufacturing, in that order. Equipment with better seals and hoses would reduce emissions and result in less need for system maintenance.[61]

When car air conditioners are serviced, it is now standard practice to drain the coolant and let it evaporate. Several companies have seen the folly of this approach and designed recovery systems, known as "vampires." The refrigerant is pumped out of the compressor, purified, and reinjected into the automobile by equipment costing several thousand dollars. Because the coolant contains few contaminants, up to 95 percent of it can be reused. Refrigerant can also be stored and transported to a central recycler, though this option appears to offer less promise.[62]

Markets for the on-site equipment include mass-transit companies, airplane manufacturers, government agencies, automobile dealerships, and high-volume service stations. The U.S. automobile industry is currently developing quality standards for recovery and recycling operations and should have the infrastructure and perhaps the equipment in place by 1989's air conditioning season. If the industry is not recycling at a satisfactory level by 1992, the EPA will make the practice mandatory.[63]

To recover CFCs from junked automobiles and other appliances years after they are produced requires either a collection system or a bounty scheme to encourage reclamation by salvagers. Several towns in West

> "Several towns in West Germany are starting to collect discarded refrigerators to keep the CFCs from reaching the atmosphere."

Germany are starting to collect discarded household refrigerators to keep the CFCs in the coolant and the insulating foam from reaching the atmosphere. The refrigerants will be recycled and the foam will be incinerated in high-temperature furnaces. Although a few other countries are contemplating this approach, most view it as economical only for large commercial and industrial units, not for the small volumes that would be recovered from household appliances.[64]

Over the longer term, phasing out the use and emissions of CFCs will require the development of chemical substitutes that do not harm the ozone layer. The challenge is to find alternatives that perform the same function for a reasonable cost, that do not require major equipment modifications, that are nontoxic to workers and consumers, and that are environmentally benign.

Petroferm, a small company in Fernandina Beach, Florida, has developed a substitute solvent called BioAct EC-7. Made with terpenes found in citrus fruit rinds, the chemical is biodegradable, nontoxic, and noncorrosive. BioAct EC-7 has been tested by AT&T at three of its plants and was found to be effective and economically competitive, even counting the cost of replacing cleaning machinery. AT&T, which used some 1,400 tons of CFC-113 in 1986, expects to replace about one-fourth of its CFC use with BioAct EC-7 over the next two years. An outside analysis estimates that the new compound could substitute for 30–55 percent of total projected CFC-113 use in the U.S. electronics industry.[65]

Du Pont and Imperial Chemical Industries (ICI), the world's two largest CFC producers, appear convinced that the replacement chemical for CFC-12 in air conditioners and refrigerators will be chlorine-free HFC-134a. Du Pont has already announced plans to build a commercial plant costing $25 million in Corpus Christi, Texas. The company says annual production will exceed 1,000 tons starting in 1990. The plant will be the fourth Du Pont facility for HFC-134a, and the seventh in the company's overall program to develop CFC alternatives. The substitute will likely sell for $2 per kilogram, some seven times the price of CFC-12.[66]

Work is also under way to develop new chemical blowing agents for both flexible and rigid foams. Union Carbide recently announced that it has found a substitute chemical for blowing the soft polyurethane foam used in furniture padding. According to the company, its new product, Ultracel, is already commercially available and could eliminate 70 percent of the CFCs employed in the flexible-foam industry. Dow Chemical, a leading manufacturer of rigid-foam insulation, has announced that it will stop using regulated CFCs by 1989.[67]

Many of the major chemical manufacturers are placing their bets on HCFC-22, -123, -141b, and -142b. The added hydrogen atom makes the ozone depletion potential of these compounds only 5 percent that of the chemicals they would replace. Their cost, on the other hand, would be three to five times greater.[68]

One major delay associated with the commercialization of new chemical compounds is the need for extensive toxicity testing; tests are run for five to seven years. To expedite this process, 14 CFC producers from Europe, Japan, Korea, and the United States have decided to pool their efforts in a multimillion-dollar joint testing program. HCFC-123 and HFC-134a are the first two chemicals chosen to undergo long-term testing. HCFC-22 has already passed. Results will be shared among members and, if promising, ought to be passed along to regulatory agencies to speed the approval process.[69]

Some alternative foam-blowing agents are already available and have been used for years. These include methylene chloride, pentane, and carbon dioxide. Although still viewed as possible substitutes, each has drawbacks. Methylene chloride is a carcinogen and difficult to dispose of, pentane is highly flammable and contributes to photochemical smog, and carbon dioxide results in denser flexible foams and rigid foams with poorer insulating properties.[70]

In some instances, new product designs can eliminate or reduce the need for CFCs and substitute chemicals while providing additional benefits. In automobiles, for instance, side vent windows, window glazings that slow solar absorption, and new solar ventilation systems can reduce interior heating and curb or eliminate the need for air conditioning, thereby saving energy. Helium refrigerators, long used

> "In some instances, new product designs can eliminate or reduce the need for CFCs while providing additional benefits."

for space and military applications, have been adapted for civilian use in trucks and homes. Cryodynamics, a New Jersey company, will soon produce 9 million helium-cooled refrigerators in Shanghai. The units use less than half the energy of conventional systems. In Japan, ammonia refrigerants are used in energy-efficient commercial buildings.[71]

Rigid-foam insulation in refrigerators and freezers may ultimately be replaced by vacuum insulation, the type used in thermos bottles. Work done at the U.S. Solar Energy Research Institute indicates that vacuum panels take up less space than foams and make appliances more energy efficient.[72]

Halon emissions appear relatively easy to curtail, although there are no promising substitutes on the horizon. Most halons produced are never used, they just need to be available in case of emergency. At present, halon flooding systems are tested when first installed by releasing all the gas in the system. Discharge testing now contributes more emissions than fire fighting does. Using alternative chemicals or other testing procedures that are acceptable to the insurance industry and eliminating accidental discharges would cut annual emissions by two-thirds.[73]

Another large source of halon emissions is fire fighter training. The U.S. military, with one of the world's largest programs, has recently introduced the use of simulators that do not require actual chemical release. ICI is establishing a recycling service for halon 1211 so that contaminated supplies, and those that would otherwise be disposed of, can be recovered.[74]

The approach taken to reducing CFC and halon emissions varies greatly among nations and industries. Companies in Sweden, for example, view the development of alternative products and processes as an economic opportunity. They are poised to seize new international markets in a changing global economy. On the other hand, the major chemical producers in France, Japan, the United Kingdom, the United States, and West Germany have traditionally viewed emissions controls as a threat to their international competitiveness. They have been loath to go along with unilateral control measures for fear of losing their market share.

The time has come to ask if the functions performed by CFCs are really necessary and, if they are, whether they can be performed in new ways. For example, must all computer chips be cleaned? For those that require cleaning, are water- or alcohol-based solvents sufficient? The U.S. Army and Navy will not allow the question to be asked. They require that electronic components be cleaned with CFCs, thereby discouraging manufacturers from exploring alternatives.[75]

If all known technical control measures were used, total CFC and halon emissions could be reduced by approximately 90 percent. Many of these control strategies are already cost-effective, and more will become so as regulations push up the price of ozone depleting chemicals. The speed with which controls are introduced will determine the extent of ozone depletion in the years ahead and when healing of the ozone layer will begin.

Beyond Montreal

An international treaty to halve the production of a ubiquitous, invisible chemical feared responsible for destroying an invisible shield is unprecedented. The achievement is a tribute to the United Nations Environment Program that spearheaded the effort, to government negotiators from all countries, especially those from the so-called Toronto group who kept pushing for tougher regulations, and to the many nongovernmental organizations and scientists who worked to build support among policymakers and the general public.

Unfortunately, for several reasons, the Montreal Protocol will not save the ozone layer. First, many inducements were offered to enhance the treaty's appeal to prospective signatories. These include extended deadlines for developing countries and centrally planned economies, allowances to accommodate industry restructuring, and loose definitions of the products that can legitimately be traded internationally. The cumulative effect of these loopholes means that, even with

> "The recognition that global warming may have already begun strengthens the case for further and more rapid reductions in CFC emissions."

widespread participation, the protocol's goal of halving worldwide CFC use by 1998 will not be met.[76]

Second, recent scientific findings show that more ozone depletion has already occurred than treaty negotiators assumed would happen in 100 years. A recent EPA report concluded that by 2075 even with 100-percent global participation in the protocol, chlorine concentrations in the atmosphere would triple. The agreement will not arrest depletion, merely slow its acceleration.[77]

Third, several chemicals not regulated under the treaty are major threats to the ozone layer. Methyl chloroform and carbon tetrachloride together contributed 13 percent of total ozone depleting chemical emissions in 1985. As the use of controlled chemicals diminishes, the contribution of these two uncontrolled compounds will grow. Methyl chloroform is widely used as a solvent, especially for metal cleaning, and carbon tetrachloride, though used primarily to manufacture CFCs in western industrial countries, is still used as a solvent in Eastern European and developing nations. Methyl chloroform emissions currently exceed those for any of the CFCs. Its short atmospheric lifetime of eight years makes it one of the few chemicals whose control would provide quick results.[78]

The recognition that global warming may have already begun strengthens the case for further and more rapid reductions in CFC emissions. CFCs currently account for 15-20 percent of the greenhouse effect and absorb wavelengths of infrared radiation that other greenhouse gases would allow to escape. Indeed, one molecule of the most widely used CFCs is as effective in trapping heat as 15,000 molecules of carbon dioxide, the most abundant greenhouse gas. Since there are only some two dozen producers of the chemicals, CFCs are the easiest greenhouse gas to control. Eliminating CFC emissions is the only way to quickly reduce the rate of temperature rise.[79]

In light of these findings, logic suggests a virtual phaseout of CFC and halon emissions by all countries as soon as possible. Releases of other chlorine- and bromine-containing compounds not currently covered under the treaty also need to be controlled and in some cases halted. According to EPA analyses, 45 percent of projected chlorine growth in the stratosphere by 2075 will stem from allowed use of controlled compounds, 40 percent will come from chlorine-containing chemicals that are not covered, and 15 percent will be from the emissions of non-participant nations.[80]

The timing of the phaseout is crucial. Analysts at EPA examined the effects of a 100-percent CFC phaseout by 1990 and a 95-percent phaseout by 1998. Peak chlorine concentrations would differ by 0.8 parts per billion, some one-third of current levels. And under the slower phasedown, atmospheric cleansing would be prolonged considerably: chlorine levels would remain higher than the peak associated with the accelerated schedule for at least 50 years.[81]

As noted, it is technically feasible to reduce CFC and halon emissions by at least 90 percent. The challenge is for governments to muster the political will to do so.

Sweden is the first country to move beyond endorsing a theoretical phaseout. In June 1988 the parliament, after extensive discussions with industry, passed legislation that includes specific deadlines for banning the use of CFCs in new products. Consumption is to be halved by 1991 and virtually eliminated by 1995. Sterilization uses and the small quantity remaining in aerosols are to be phased out by the end of 1988. Use in packaging materials is to cease a year later. CFCs used as an engineering solvent and for blowing flexible and extruded polystyrene foams are to be discontinued by 1991. Blow molding of rigid foams, dry-cleaning, and coolant uses are to cease by the end of 1994 at the latest. Under no circumstance may CFCs be replaced with chemicals that pose environmental or health hazards.[82]

If it becomes possible to phase out any of these uses sooner, Swedish industries will be required to do so. In the interim, the Swedish government plans to offer incentives and provide financial support for the research and development of recovery and recycling technologies, of alternative products, and of means to keep discarded CFCs from reaching the atmosphere. The latter includes collection systems for coolants and incineration technologies for rigid foams. Sweden is currently responsible for less than 1 percent of global CFC use,

> "Environmental agencies in the United Kingdom, the United States, and West Germany have endorsed emissions reductions of at least 85 percent."

however, so its approach will have to be adopted by many more countries before a significant dent is made in global emissions.[83]

Broader support for the concept of a phaseout began to emerge following the release of the NASA Ozone Trends Panel report. Du Pont, the world's largest CFC manufacturer, was the first to commit itself to ending the production of all regulated CFCs by the year 2000. Other companies, including Allied Signal, ICI, and Pennwalt, soon endorsed similar plans, but most failed to include a schedule for abandoning chemical production.[84]

Industries that use CFCs to make consumer products have been slower to respond. Aerosol producers and users of foam packaging, who have ready access to alternatives and who have been pressured by consumers, have been the most willing to abandon CFCs. Manufacturers of air conditioning and refrigeration equipment, on the other hand, are reluctant to engage in costly retooling. For them, CFCs represent a very small percentage of production costs, and they appear ready to pay the higher price that will be commanded for regulated chemicals and for substitutes. Unfortunately, one of the most beneficial uses of CFCs—in energy-saving insulation foams—may be priced out of the market.

Environmental agencies in the United Kingdom, the United States, and West Germany have endorsed emissions reductions of at least 85 percent. This represents a significant turnaround in the United Kingdom and West Germany, and a resumption of long-held policy in the United States. Chemical producers in these three countries account for over half the global output of controlled substances. Given the clout of these governments in international economic and diplomatic circles, they possess considerable leverage to influence more recalcitrant nations.[85]

Levying a tax on newly manufactured CFCs and other ozone depleting substances is one way governments can cut emissions and accelerate the adoption of alternative chemicals and technologies. If the tax increased in step with mandatory production cutbacks, it would eliminate windfall profits for producers, encourage recovery and recycling processes, stimulate use of new chemicals, and provide a source of funding for new technologies and for needed research. Encouraging investments in recycling networks, incinerators for rigid foams, and collection systems for chemicals that would otherwise be discarded could substantially trim emissions from existing products, from servicing operations, and from new production runs.[86]

Although chemical manufacturers are spending some $100 million annually to develop safe chemical substitutes, they have no interest in alternative product designs that would cut into their markets. Research on new refrigeration, air conditioning, and insulation processes is the most worthy of government support. Phasing out the use of CFCs in these applications would protect the ozone layer and delay the greenhouse effect—directly by reducing CFC emissions and indirectly by promoting more energy-efficient technologies that would trim carbon dioxide emissions. Cooling cars, offices, and factories with equipment powered by fuels and dependent on chemicals that warm the earth is ludicrous. Unfortunately, international funding for developing such technologies now totals less than $5 million.[87]

As mentioned in the text of the Montreal Protocol, results of this research, as well as new technologies and processes, need to be shared with developing countries. Ozone depletion and climate warming are undeniably global in scope. Not sharing information on the most recent developments is like refusing to tell the driver of a car that is about to hit you where the brakes are. And it ensures that environmentally damaging and outdated equipment will continue to be used for years to come, further eroding the Third World technology base.

Developing countries are an important part of the control strategy because of their large and growing populations and their rapidly increasing CFC use. Some of the key developing countries are Brazil, China, India, Indonesia, and South Korea. In China, for example, only 1 household out of 10 now owns a refrigerator, but the government hopes that by the year 2000 every kitchen will have one. South Korea and Brazil are fast becoming major players in the world automobile market.[88]

Under the Montreal Protocol, a scientific assessment of current ozone depletion is scheduled to occur from April to August 1989. This is to be followed in April 1990 by a meeting of treaty negotiators to consider the evidence and decide what further actions are called for. Given recent developments, Mostafa Tolba, executive director of the United Nations Environment Program and a prime mover behind the treaty, has been widely petitioned to expedite the process and is in favor of doing so. Indeed, he had earlier pledged to reopen the agreement if it turned out that the Antarctic hole was caused by CFCs. The United Kingdom, the United States, West Germany, and possibly Japan and the Soviet Union appear amenable to taking faster action. France is the only major producer country still dragging its feet.[89]

The scientific fundamentals of ozone depletion and climate change are known, and there is widespread agreement that both have already begun. Although current models of future change vary in their predictions, the evidence is clear enough to warrant an immediate response. Because valuable time was lost when governments and industries relaxed their regulatory and research efforts during the early eighties, a crash program is now essential. Human health, food supplies, and the global climate all hinge on the support that can be garnered for putting an end to chlorine and bromine emissions.

Notes

1. Joseph C. Farman et al., "Large Losses of Total Ozone in Antarctica Reveal Seasonal ClO$_x$/NO$_x$ Interaction," *Nature*, May 16, 1985; Paul Brodeur, "Annals of Chemistry: In the Face of Doubt," *New Yorker*, June 9, 1986.

2. Subcommittee on Environmental Protection and Subcommittee on Hazardous Wastes and Toxic Substances, *Implications of the Findings of the Expedition to Investigate the Ozone Hole over the Antarctic*, Committee on Environment and Public Works, U.S. Senate, October 27, 1987.

3. National Aeronautics and Space Administration (NASA), "Executive Summary of the Ozone Trends Panel," Washington, D.C., March 15, 1988.

4. United Nations Environment Program (UNEP), "Montreal Protocol on Substances that Deplete the Ozone Layer," 1987; country updates from United Nations (UN) Treaty Office, New York; Office of Technology Assessment (OTA), U.S. Congress, "An Analysis of the Montreal Protocol on Substances that Deplete the Ozone Layer," Washington, D.C., December 10, 1987 (rev. February 1, 1988); NASA, "Ozone Trends"; John S. Hoffman and Michael J. Gibbs, *Future Concentrations of Stratospheric Chlorine and Bromine* (Washington, D.C.: U.S. Environmental Protection Agency (EPA), 1988).

5. James E. Hansen, NASA Goddard Institute for Space Studies, Testimony before the Committtee on Energy and Natural Resources, U.S. Senate, June 23, 1988; Linda J. Fisher, EPA, Testimony before the Subcommittee on Energy and Power, Committee on Energy and Commerce, U.S. House of Representatives, September 22, 1988; T.M.L. Wigley, "Future CFC Concentrations under the Montreal Protocol and Their Greenhouse-effect Implications," *Nature*, September 22, 1988.

6. Farman et al., "Large Losses"; Mario Molina and F. Sherwood Rowland, "Stratospheric Sink for Chlorofluoromethanes: Chlorine Atom Catalyzed Destruction of Ozone," *Nature*, June 28, 1974.

7. Douglas G. Cogan, *Stones in a Glass House: CFCs and Ozone Depletion* (Washington, D.C.: Investor Responsibility Research Center, 1988); Richard S. Stolarski, "The Antarctic Ozone Hole," *Scientific American*, January 1988.

8. Susan Soloman, National Oceanic and Atmospheric Administration (NOAA), interview on "The Hole in the Sky," *NOVA*, WGBH-Boston, February 24, 1987.

9. Stolarski, "The Antarctic Ozone Hole"; "Airborne Antarctic Ozone Experiment," NASA, Washington, D.C., July 1987; Shirley Christian, "Pilots Fly over the Pole into the Heart of Ozone Mystery," *New York Times*, September 22, 1987.

10. Mario J. Molina et al., "Antarctic Stratospheric Chemistry of Chlorine Nitrate, Hydrogen Chloride, and Ice: Release of Active Chlorine," *Science*, November 27, 1987; Mario Molina, "The Antarctic Ozone Hole," *Oceanus*, Summer 1988; F. Sherwood Rowland, University of California at Irvine, Testimony before the Committee on Environment and Public Works, U.S. Senate, September 14, 1988.

11. James G. Anderson, Harvard University, Testimony before the Environmental Protection and the Hazardous Wastes and Toxic Substances Subcommittees, Committee on Environment and Public Works, U.S. Senate, October 27, 1987.

12. Molina, "The Antarctic Ozone Hole"; NASA, "Ozone Trends."

13. Michael McElroy, "The Challenge of Global Change," *New Scientist*, July 28, 1988; Donald R. Blake and F. Sherwood Rowland, "Continuing Worldwide Increase in Tropospheric Methane, 1978 to 1987," *Nature*, March 4, 1988.

14. Pamela S. Zurer, "Studies on Ozone Destruction Expand Beyond Antarctic," *Chemical and Engineering News*, May 30, 1988.

15. Ibid.; Malcolm W. Browne, "New Ozone Threat: Scientists Fear Layer is Eroding at North Pole," *New York Times*, October 11, 1988.

16. Zurer, "Studies on Ozone Destruction"; Molina, "The Antarctic Ozone Hole"; John Gribben, "Satellite Failure Threatens Ozone Probe," *New Scientist*, July 14, 1988.

17. Rowland, Testimony; R. Monastersky, "Arctic Ozone: Signs of Chemical Destruction," *Science News*, June 11, 1988; Robert T. Watson, NASA, Testimony before the Environmental Protection and the Hazardous Wastes and Toxic Substances Subcommittees, Committee on Environment and Public Works, U.S. Senate, October 27, 1987.

18. NASA, "Ozone Trends."

19. Ibid.; McElroy, "Challenge of Global Change."

20. Robert T. Watson, "Present State of Knowledge of the Ozone Layer," presented to The Changing Atmosphere: Implications for Global Security, Toronto, June 27–30, 1988; Rowland, Testimony.

21. James Gleik, "Even with Action Today, Ozone Loss Will Increase," *New York Times*, March 20, 1988.

22. NASA, "Ozone Trends"; EPA, *Regulatory Impact Analysis: Protection of Stratospheric Ozone*, Vol. I (Washington, D.C.: 1987).

23. J.C. van der Leun, "Health Effects of Ultraviolet Radiation," draft report to the UNEP Coordinating Committee on the Ozone Layer, Effects of Stratospheric Modification and Climate Change, Bilthoven, Netherlands, November 19–21, 1986 (hereinafter cited as UNEP Coordinating Committee).

24. Ibid.; Paul Strickland et al., "Sunlight, Ozone, and Skin Cancer," *Health & Environment Digest*, May 1988; "Effects of Ozone Layer Modification," UNEP Coordinating Committee; EPA, *Regulatory Impact Analysis*; Polly Penhale, National Science Foundation, Washington, D.C., private communication, September 28, 1988

25. National Cancer Institute, *1987 Annual Cancer Statistics Review: Including Cancer Trends: 1950–1985* (Bethesda, Maryland: National Institutes of Health, 1987); Arjun Makhijani et al., *Saving Our Skins: Technical Potential and Policies for the Elimination of Ozone Depleting Chlorine Compounds* (Washington, D.C.: Environmental Policy Institute/Institute for Energy and Environmental Research, 1988); EPA, *Regulatory Impact Analysis*.

26. Robin Russell Jones, "Ozone Depletion and Cancer Risk," *Lancet*, August 22, 1987; Janice Longstreth, "Health Effects Associated with Stratospheric Ozone Depletion," in *The Sky Is the Limit*, Dr. Karola Taschner, editor, (Brussels: European Environmental Bureau, 1987); Darrel Rigel, New York University Medical Center, Testimony before the Subcommittee on Health and the Environment, Committee on Energy and Commerce, U.S. House of Representatives, March 9, 1987.

27. EPA, *Regulatory Impact Analysis*.

28. Margaret Kripke, M.D., Anderson Hospital and Tumor Institute, Testimony before the Environmental Protection and the Hazardous Wastes and Toxic Substances Subcommittees, Committee on Environment and Public Works, U.S. Senate, May 12–14, 1987; EPA, *Regulatory Impact Analysis*; Janice Longstreth, ICF Inc., Washington, D.C., private communication, September 28, 1988.

29. EPA, *Regulatory Impact Analysis*; Longstreth, private communication; "Chlorofluorocarbons: A Valuable Chemical Threatens the Atmosphere," *Health & Environment Digest*, May 1988.

30. "Risks to Crops and Terrestrial Ecosystems From Enhanced UV-B Radiation," draft report to the UNEP Coordinating Committee; Alan Teramura, "The Potential Consequences of Ozone Depletion Upon Global Agriculture," in J. Titus, ed., *Effects of Changes in Stratospheric Ozone and Global Climate*

(Washington, D.C.: EPA, 1986); Alan H. Teramura and N.S. Murali, "Intraspecific Differences in Growth and Yield of Soybean Exposed to Ultraviolet-B Radiation Under Greenhouse and Field Conditions," *Environmental and Experimental Botany*, Vol. 26, No. 1, 1986.

31. James Falco, director, Office of Environmental Processes and Effects Research, EPA, Testimony before the Subcommittee on Natural Resources, Agriculture Research and Environment, Committee on Science, Space and Technology, U.S. House of Representatives, March 10 and 12, 1987.

32. Robert C. Worrest, "What Are the Effects of UV-B Radiation on Marine Organisms?" Testimony before the West German Bundestag, Commission on Preventive Measures to Protect the Earth's Atmosphere, April 27, 1988.

33. Robert C. Worrest, "Solar Ultraviolet-B Radiation Effects on Aquatic Organisms," draft report to the UNEP Coordinating Committee; "Dinosaurs Doomed by a Dearth of Plankton," *New Scientist*, March 17, 1988.

34. Worrest, "Solar Ultraviolet-B Radiation Effects."

35. Office of Air and Radiation, *Assessing the Risks of Trace Gases that Can Modify the Stratosphere* (Washington, D.C.: EPA, 1987).

36. Philip Shabecoff, "Ozone Pollution is Found at Peak in Summer Heat," *New York Times*, July 31, 1988; Harold Dovland, "Monitoring European Transboundary Air Pollution," *Environment*, December 1987; EPA, *Regulatory Impact Analysis*.

37. Private communications with officials at EPA, National Science Foundation, U.S. Department of Agriculture, and National Institutes of Health; Hartmut Keune, Ecological Research Division, West German Ministry for Research and Technology, Bonn, private communication, June 28, 1988.

38. EPA, *Regulatory Impact Analysis*.

39. Cogan, *Stones in a Glass House*.

40. Chemical Manufacturers Association (CMA), "Production, Sales, and Calculated Release of CFC-11 and CFC-12 Through 1986," Washington, D.C., November 18, 1987.

41. Michael Weisskopf, "CFCs: Rise and Fall of Chemical 'Miracle'," *Washington Post*, April 10, 1988; EPA, Addenda to *Regulatory Impact Analysis*; *The Montreal Protocol: A Briefing Book* (Rosslyn, Virginia: Alliance for Responsible CFC Policy, 1987).

42. Steve Risotto, Halogenated Solvents Industry Alliance, Washington, D.C, private communication, August 31, 1988.

43. Cogan, *Stones in a Glass House*.

44. Ron Wolf, "Ozone Layer Negotiations Target Chlorofluorocarbons," *Journal of Commerce*, August 13, 1987; Alliance for Responsible CFC Policy, *The Montreal Protocol*.

45. Cogan, *Stones in a Glass House*; P.H. Gamlen et al., "The Production and Release to the Atmosphere of CCl_3F and CCl_2F_2 (Chlorofluorocarbons CFC-11 and CFC-12)," *Atmospheric Environment*, Vol. 20, No 6, 1986; Elizabeth Festa Gormley, Chemical Manufacturers Association, Washington, D.C., private communication, August 31, 1988; Christopher F.P. Bevington, Metro Consulting Group, Ltd., London, private communication, May 23, 1988; Risotto, private communication.

46. EPA, *Regulatory Impact Analysis*; Richard Monastersky, "Decline of the CFC Empire," *Science News*, April 9, 1988; House of Commons Environment Committee, *Air Pollution* (London: Her Majesty's Stationery Office, 1988); Cogan, *Stones in a Glass House*; "White Paper on the Environment in Japan 1988," Japanese Environment Agency, May 1988.

47. CMA, "Production, Sales, and Calculated Release."

48. Cogan, *Stones in a Glass House*.

49. CMA, "Production, Sales, and Calculated Release"; Risotto, private communication.

50. Michael Weisskopf, "EPA Urges Halt in Use of CFCs," *Washington Post*, September 27, 1988; Hoffman and Gibbs, *Future Concentrations*.

51. UNEP, "Montreal Protocol"; UN Treaty Office.

52. Cogan, *Stones in a Glass House*; CMA, "Production, Sales, and Calculated Release"; Michael Kavanaugh et al., "An Analysis of the Economic Effects of Regulatory and Non-regulatory Events Related to the Abandonment of Chlorofluorocarbons as Aerosol Propellants in the United States From 1970 to 1980, with a Discussion of Applicability of the Analysis to Other Nations," ICF Inc., Washington, D.C., February 1986 (rev.); Nigel Haigh, *EEC Environmental Policy & Britain* (Harlow, Essex: Longman Group UK, Ltd., 1987).

53. Haigh, *EEC Environmental Policy & Britain*; cumulative emissions reductions from EPA, *Regulatory Impact Analysis*; CMA, "Production, Sales, and Calculated Release."

54. CMA, "Production, Sales, and Calculated Release."

55. "The Aerosol Industry and CFCs: A Parting of the Ways," *ENDS Report*, January 1988; Mark Vandenreeck, Belgian Embassy, Washington, D.C., private communication, April 14, 1988; Wolf Dieter Garber, Umweltbundesamt, Berlin, West Germany, private communication, June 24, 1988; "Aerosol Makers to Offer Voluntary Labeling for Products Without CFCs, Association Says," *International Environment Reporter*, June 8, 1988; Vera Rich, "Growing Reaction to Ozone Hole in Soviet Union," *Nature*, August 25, 1988.

56. F. Camm et al., "The Social Cost of Technical Control Options to Reduce Emissions of Potential Ozone Depleters in the United States: An Update," Rand Corporation, Santa Monica, California, May 1986; Alan S. Miller and Irving M. Mintzer, "The Sky Is the Limit: Strategies for Protecting the Ozone Layer," World Resources Institute, Washington, D.C., November 1986.

57. M. Drechsler, Umweltbundesamt, Berlin, West Germany, private communication, June 24, 1988; L.R. Wallace, AT&T, Princeton, N.J., private communication, March 9, 1988; Maurice Verhille, Atochem, Paris, private communication, June 17, 1988; Kevin Fay, Alliance for Responsible CFC Policy, Rosslyn, Virginia, private communication, October 14, 1988; J. Rodgers, Allied-Signal, "Recycling and Recovery of Solvents in the Electronics Industry," in "Proceedings of Conference and Trade Fair: Substitutes and Alternatives to CFCs and Halons," EPA, Washington, D.C., January 13–15, 1988 (hereinafter cited as Substitutes and Alternatives Conference).

58. "Takeshita Cabinet Approves Ozone Bill Including Tax Incentives for CFC Recycling," *International Environment Reporter*, April 13, 1988.

59. C.H. Mueller, "Report on Realization and Results with a Full Scale CFC-11 Recovery Unit in the Flexible Foam Slabstock Production at Recticel in Kesteren Holland," Escher Hoogezand, Netherlands, March 11, 1987; Dr. H. Creyf, Recticel, Testimony before the West German Bundestag, Commission on Preventive Measures to Protect the Earth's Atmosphere, April 13, 1988; National Swedish Environmental Protection Board, *CFCs/Freons: Proposals to Protect the Ozone Layer* (Solna, Sweden: 1987); "Foam Plastics: Next in Line for the CFCs Campaign," *ENDS Report*, March 1988; N.C. Vreenegoor, "Environmental Considerations in the Production of Flexible Slabstock," Substitutes and Alternatives Conference.

60. Mueller, "Report on Realization and Results with a Full Scale CFC-11 Recovery Unit"; Creyf, Testimony; National Swedish Environmental Protection Board, *CFCs/Freons*; "Foam Plastics: Next in Line," *ENDS Report*; Vreenegoor, "Environmental Considerations in the Production of Flexible Slabstock."

61. EPA, Addenda to *Regulatory Impact Analysis*.

62. Sarah L. Clark, "Protecting the Ozone Layer: What You Can Do," Environmental Defense Fund, 1988; Jean Lupinacci, EPA, private communication, October 14, 1988; Kenneth Manz, Robinair, Montpelier, Ohio, private communication, August 11, 1988.

63. Clark, "Protecting the Ozone Layer"; Lupinacci, private communication.

64. Mr. Pautz, Umweltbundesamt, Berlin, West Germany, private communication, June 24, 1988.

65. John R. Fisher, "A New Rosin Defluxing Alternative," AT&T, Princeton, N.J., 1988; Philip Shabecoff, "New Compound Is Hailed as Boon to Ozone Shield," *New York Times*, January 14, 1988; Pamela S. Zurer, "Search Intensifies for Alternatives to Ozone Depleting Halocarbons," *Chemical and Engineering News*, February 8, 1988; Sudhakar Kesavan, "Overview of CFC-113 Use in the Electronics Industry and Control Options Available," Substitutes and Alternatives Conference.

66. Laurie Hays, "Du Pont Plans Plant to Produce Refrigerant Harmless to Ozone," *Wall Street Journal*, September 30, 1988; "Du Pont Plans Commercial-scale Plant for Production of CFC-12 Substitute," *Journal of Commerce*, September 30, 1988; Malcolm Gladwell, "Du Pont Plans to Make CFC Alternative," *Washington Post*, September 30, 1988.

67. "Carbide Easing an Ozone Peril," *Washington Post*, August 6, 1988; "Dow to Curtail CFCs," *Washington Post*, May 14, 1988.

68. Thomas P. Nelson, "Findings of the Chlorofluorocarbon Chemical Substitutes International Committee," EPA, Washington, D.C., 1988; Meirion Jones, "In Search of Safe CFCs," *New Scientist*, May 26, 1988; U. Bohr, Du Pont, Testimony before the West German Bundestag, Commission on Preventive Measures to Protect the Earth's Atmosphere, April 25, 1988; Imperial Chemical Industries (ICI), Testimony before the West German Bundestag, Commission on Preventive Measures to Protect the Earth's Atmosphere, April 29, 1988.

69. "Du Pont Sees Progress in Replacing Fluorocarbons," *Chemical Marketing Reporter*, January 11, 1988; "Korean Firm Joins in International Effort to Pool Knowledge on CFC Toxicity Testing Studies," *International Environmental Reporter*, April 13, 1988; Greg Freiherr, "Can Chemists Save the World from Chemists?" *The Scientist*, May 16, 1988.

70. Camm et al., "Social Cost of Technical Control Options"; National Swedish Environmental Protection Board, *CFCs/Freons*.

71. Makhijani et al., *Saving Our Skins*; Nick Sundt, OTA, Washington, D.C., private communication, July 28, 1988.

72. Tom Potter, "Potential for Offsetting CFCs with Advanced Insulation," Substitutes and Alternatives Conference.

73. John W. Mossel, "Uses of Halons and Opportunities for Emission Reductions: Size and Structure of the Market," Substitutes and Alternatives Conference.

74. Ibid.; Tom Moorehouse, "The Air Force Halon/Ozone Research Program," Substitutes and Alternatives Conference.

75. Gregory C. Munie, AT&T Bell Laboratories, "Experience with the Use of Aqueous Cleaning in the Electronics Industry," Substitutes and Alternatives Conference; Leo Lambert, "Digital Equipment Corporation Experience with Aqueous Cleaning," Substitutes and Alternatives Conference; Eileen B. Claussen, "Moving Forward Together," *The Environmental Forum*, July/August 1988; Kathi Johnson, U.S. Navy, "Alternative Cleaning Materials: Research Topics for the Military," Substitutes and Alternatives Conference.

76. UNEP, "Montreal Protocol"; OTA, "An Analysis of the Montreal Protocol."

77. Hoffman and Gibbs, *Future Concentrations*.

78. James K. Hammitt et al., "Future Emission Scenarios for Chemicals that May Deplete Stratospheric Ozone," *Nature*, December 24, 1987; Arjun Makhijani, Institute for Energy and Environmental Research, Takoma Park, Maryland, private communication, October 26, 1988.

79. Hansen, Testimony before the Committee on Energy and Natural Resources; Fisher, Testimony before the Subcommittee on Energy and Power; Wigley, "Future CFC Concentrations"; F. Sherwood Rowland and Daniel G. Aldrich Jr., "Chlorofluorocarbons, Stratospheric Ozone, and the Antarctic 'Ozone Hole'," *Environmental Conservation*, Summer 1988.

80. Hoffman and Gibbs, *Future Concentrations*.

81. Ibid.

82. Government of Sweden, "Environmental Policy for the 1990s," Environmental Bill, March 4, 1988; Sverker Hogberg, scientific counselor, Embassy of Sweden, Washington, D.C., private communication, October 13, 1988.

83. National Swedish Environmental Protection Board, *CFCs/Freons*; Government of Sweden, "Environmental Policy for the 1990s."

84. Liz Cook, Friends of the Earth, Washington, D.C., private communication, October 6, 1988.

85. Lord Caithness, UK minister of the environment, statement to the press, October 3, 1988; Philip Shabecoff, "EPA Chief Asks Total Ban on Ozone-harming Chemicals," *Washington Post*, September 27, 1988; Wolf Dieter Garber, private communication.

86. For further discussion of a tax on ozone depleting substances see Cogan, *Stones in a Glass House*, and Makhijani, *Saving Our Skins*.

87. Mark Ledbetter, American Council for an Energy Efficient Economy, Washington, D.C., private communication, October 12, 1988; Terry Statt, U.S. Department of Energy, Washington, D.C., private communication, October 12, 1988.

88. National Swedish Environmental Protection Board, *CFCs/Freons*.

89. Joan Martin-Brown, UNEP, Washington, D.C., private communication, November 1, 1988; Senator John H. Chafee, Statement in *Preparing for Climate Change: Proceedings of the First North American Conference on Preparing for Climate Change: A Cooperative Approach* (Washington, D.C.: Government Institutes, Inc., 1988).

CYNTHIA POLLOCK SHEA is a Senior Researcher with the Worldwatch Institute, and coauthor of *State of the World 1988*. Her research deals with ozone depletion and energy and waste management technologies and policies. She is a graduate of the University of Wisconsin–Madison, where she studied economics, environmental studies, and international relations.

THE WORLDWATCH PAPER SERIES

No. of Copies*

- 1. **The Other Energy Crisis: Firewood** by Erik Eckholm.
- 3. **Women in Politics: A Global Review** by Kathleen Newland.
- 7. **The Unfinished Assignment: Equal Education for Women** by Patricia L. McGrath.
- 10. **Health: The Family Planning Factor** by Erik Eckholm and Kathleen Newland.
- 16. **Women and Population Growth: Choice Beyond Childbearing** by Kathleen Newland.
- 18. **Cutting Tobacco's Toll** by Erik Eckholm.
- 21. **Soft Technologies, Hard Choices** by Colin Norman.
- 25. **Worker Participation—Productivity and the Quality of Work Life** by Bruce Stokes.
- 28. **Global Employment and Economic Justice: The Policy Challenge** by Kathleen Newland.
- 29. **Resource Trends and Population Policy: A Time for Reassessment** by Lester R. Brown.
- 30. **The Dispossessed of the Earth: Land Reform and Sustainable Development** by Erik Eckholm.
- 31. **Knowledge and Power: The Global Research and Development Budget** by Colin Norman.
- 33. **International Migration: The Search for Work** by Kathleen Newland.
- 34. **Inflation: The Rising Cost of Living on a Small Planet** by Robert Fuller.
- 35. **Food or Fuel: New Competition for the World's Cropland** by Lester R. Brown.
- 36. **The Future of Synthetic Materials: The Petroleum Connection** by Christopher Flavin.
- 38. **City Limits: Emerging Constraints on Urban Growth** by Kathleen Newland.
- 39. **Microelectronics at Work: Productivity and Jobs in the World Economy** by Colin Norman.
- 40. **Energy and Architecture: The Solar Conservation Potential** by Christopher Flavin.
- 41. **Men and Family Planning** by Bruce Stokes.
- 42. **Wood: An Ancient Fuel with a New Future** by Nigel Smith.
- 43. **Refugees: The New International Politics of Displacement** by Kathleen Newland.
- 44. **Rivers of Energy: The Hydropower Potential** by Daniel Deudney.
- 45. **Wind Power: A Turning Point** by Christopher Flavin.
- 46. **Global Housing Prospects: The Resource Constraints** by Bruce Stokes.
- 47. **Infant Mortality and the Health of Societies** by Kathleen Newland.
- 48. **Six Steps to a Sustainable Society** by Lester R. Brown and Pamela Shaw.
- 49. **Productivity: The New Economic Context** by Kathleen Newland.
- 50. **Space: The High Frontier in Perspective** by Daniel Deudney.
- 51. **U.S. and Soviet Agriculture: The Shifting Balance of Power** by Lester R. Brown.
- 52. **Electricity from Sunlight: The Future of Photovoltaics** by Christopher Flavin.
- 53. **Population Policies for a New Economic Era** by Lester R. Brown.
- 54. **Promoting Population Stabilization** by Judith Jacobsen.
- 55. **Whole Earth Security: A Geopolitics of Peace** by Daniel Deudney.
- 56. **Materials Recycling: The Virtue of Necessity** by William U. Chandler.
- 57. **Nuclear Power: The Market Test** by Christopher Flavin.
- 58. **Air Pollution, Acid Rain, and the Future of Forests** by Sandra Postel.
- 59. **Improving World Health: A Least Cost Strategy** by William U. Chandler.

*Worldwatch Papers 2, 4, 5, 6, 8, 9, 11, 12, 13, 14, 15, 17, 19, 20, 22, 23, 24, 26, 27, 32, and 37 are out of print.

_____ 60. **Soil Erosion: Quiet Crisis in the World Economy** by Lester R. Brown and Edward C. Wolf.
_____ 61. **Electricity's Future: The Shift to Efficiency and Small-Scale Power** by Christopher Flavin.
_____ 62. **Water: Rethinking Management in an Age of Scarcity** by Sandra Postel.
_____ 63. **Energy Productivity: Key to Environmental Protection and Economic Progress** by William U. Chandler.
_____ 64. **Investing in Children** by William U. Chandler.
_____ 65. **Reversing Africa's Decline** by Lester R. Brown and Edward C. Wolf.
_____ 66. **World Oil: Coping With the Dangers of Success** by Christopher Flavin.
_____ 67. **Conserving Water: The Untapped Alternative** by Sandra Postel.
_____ 68. **Banishing Tobacco** by William U. Chandler.
_____ 69. **Decommissioning: Nuclear Power's Missing Link** by Cynthia Pollock.
_____ 70. **Electricity For A Developing World** by Christopher Flavin.
_____ 71. **Altering the Earth's Chemistry: Assessing the Risks** by Sandra Postel.
_____ 72. **The Changing Role of the Market in National Economies** by William U. Chandler.
_____ 73. **Beyond the Green Revolution: New Approaches for Third World Agriculture** by Edward C. Wolf.
_____ 74. **Our Demographically Divided World** by Lester R. Brown and Jodi L. Jacobson.
_____ 75. **Reassessing Nuclear Power: The Fallout From Chernobyl** by Christopher Flavin.
_____ 76. **Mining Urban Wastes: The Potential for Recycling** by Cynthia Pollock.
_____ 77. **The Future of Urbanization: Facing the Ecological and Economic Constraints** by Lester R. Brown and Jodi L. Jacobson.
_____ 78. **On the Brink of Extinction: Conserving The Diversity of Life** by Edw. Wolf.
_____ 79. **Defusing the Toxics Threat: Controlling Pesticides and Industrial W** by Sandra Postel.
_____ 80. **Planning the Global Family** by Jodi L. Jacobson.
_____ 81. **Renewable Energy: Today's Contribution, Tomorrow's Promise** by Cynth Pollock Shea.
_____ 82. **Building on Success: The Age of Energy Efficiency** by Christopher Flavi and Alan B. Durning.
_____ 83. **Reforesting the Earth** by Sandra Postel and Lori Heise.
_____ 84. **Rethinking the Role of the Automobile** by Michael Renner.
_____ 85. **The Changing World Food Prospect: The Nineties and Beyond** by Lester R. Brown.
_____ 86. **Environmental Refugees: A Yardstick of Habitability** by Jodi L. Jacobson.
_____ 87. **Protecting Life on Earth: Steps to Save the Ozone Layer** by Cynthia Pollock Shea.

_____ **Total Copies**

Bulk Copies (any combination of titles)   **Single Copy** $4.00
    2–5: $3.00 each   6–20: $2.00 each   21 or more: $1.00 each

Calendar Year Subscription (1989 subscription begins with Paper 88)   U.S. $25.00 _____

Make check payable to Worldwatch Institute
1776 Massachusetts Avenue, N.W., Washington, D.C. 20036 USA

                                     Enclosed is my check for U.S. $ _____

name

address

city   **state**   **zip/country**